Manoel Eduardo Ribas Vianna

Mathematical modelling Use of public transparency

AF665279

Manoel Eduardo Ribas Vianna

Mathematical modelling Use of public transparency

Games in the classroom

ScienciaScripts

Imprint

Any brand names and product names mentioned in this book are subject to trademark, brand or patent protection and are trademarks or registered trademarks of their respective holders. The use of brand names, product names, common names, trade names, product descriptions etc. even without a particular marking in this work is in no way to be construed to mean that such names may be regarded as unrestricted in respect of trademark and brand protection legislation and could thus be used by anyone.

Cover image: www.ingimage.com

This book is a translation from the original published under ISBN 978-613-9-63871-0.

Publisher:
Sciencia Scripts
is a trademark of
Dodo Books Indian Ocean Ltd. and OmniScriptum S.R.L publishing group

120 High Road, East Finchley, London, N2 9ED, United Kingdom
Str. Armeneasca 28/1, office 1, Chisinau MD-2012, Republic of Moldova, Europe
Printed at: see last page
ISBN: 978-620-7-71690-6

Copyright © Manoel Eduardo Ribas Vianna
Copyright © 2024 Dodo Books Indian Ocean Ltd. and OmniScriptum S.R.L publishing group

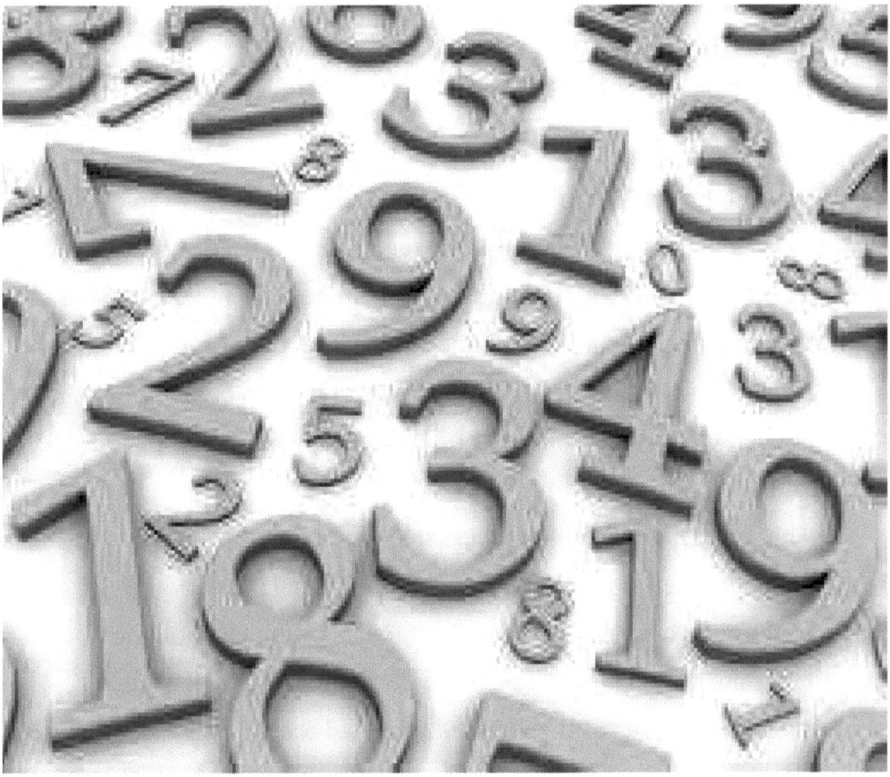

To humanity in God.

PRESENTATION

This book follows the idea of the publisher Novas Edições Acadêmicas to print the final work presented by me, in the form of an academic article, and published by the Centro Universitário Internacional de Curitiba - UNINTER in the journal Intersaberes, for the conclusion of the postgraduate course in Physics and Maths Teaching Methodology.

It deals with a subject that contributes greatly to the national context, when we discover that it's not enough to wish and hope, but to participate and build the world we want.

This book helps maths teachers to develop lessons using the mathematical modelling method, and also to understand why it is necessary to evolve or move forward in the use of the technological method that the moment demands in the formation of the new citizen, without losing the basis of the old mathematical idea.

Teaching maths isn't just about asking students to solve complicated problems, it's about giving them the thoughts, methods and logical reasoning, as well as providing them with the ability to deal with exact science and giving some the opportunity to choose to follow this marvellous universal path and providing them with the possibility of participating in a society that tends to improve professional skills.

This book, an innovator in teaching ideas by encouraging the fearless use of the advances in technology that have been created for this purpose, recalls the history of maths teaching and enables the reader to understand, explaining why it is necessary to move forward with the use of current resources. It shows the step-by-step process of setting up games or modelling work and provides examples that will make it easier for the teacher to work not just with small inspectors, but with curious people with unlimited interests, who, already technological heirs, will certainly have the intention of continuing the human legacy on planet Earth.

The use of public transparency in group mathematical modelling games accesses the student's area of experience. It can then deal with issues that they are witnessing in their daily lives, that they post on their communication devices, and that they comment on and make critical jokes about.

It should also deal with solving the problems they identify in the family reality in which their parents live, in relation to their chosen ones.

Classrooms are places of common experience, where people learn, live and are taught respect and social interaction, or group participation. Maths is a science that develops students' attention, sense of ability and security.

SUMMARY

CHAPTER 1

INTRODUCTION

Over time, human knowledge has undergone major transformations and the evolution of our scientific spirit is the result of these changes and the increase in technology. Education is firmly linked to these changes, especially maths, whose roots are intertwined with the origins of human history.

New forms of educational support, such as the use of new technologies in education or didactic transposition, whether in the art of transforming and transmitting knowledge or creating knowledge, have evolved to the point of creating paradigms regarding their use or functionality. It would be easier to ask ourselves if we are understanding the idea and doing our homework correctly.

Well-educated and capable people are the ideal for the individuals who make up a population, and they are also the intention of all nations seeking to evolve.

The National System for the Evaluation of Basic Education - SAEB, at http://portal.inep.gov.br/educacao-basica/saeb, produces information on the reality of Brazilian education, linked to the National Institute for Studies and Research - INEP, which can be accessed at http://www.inep.gov.br/, classifies the stage of development of a nation as follows:

o Very critical level: They don't identify an addition or subtraction operation involved in the problem or don't know the meaning of simple geometric figures.

o Critical level: they develop some elementary problem interpretation skills, recognise something in geometric representation, calculate areas, and solve everyday problems including small amounts of money.

o Intermediate level: Solving everyday problems involving the addition of rational numbers to two decimal places, calculating the sum and subtraction of up to three digits, including the use of a one-digit multiplication operation.

o Adequate level: they solve problems competently, complete operations with rational numbers in the four operations, notions of proportionality, including the monetary system and division calculations by two-digit numbers, including remainders.

According to the survey, only one in four Brazilians has fully mastered the skills of reading, writing and maths.

Several PCNs propose mathematical modelling in the didactic transposition of mathematics teaching (Brasil 1997, Brasil 1998, Brasil 2002), as well as the use of new support technologies, with the aim of benefiting learning and training new citizens, people capable of maintaining and evolving the society in which they live. As a suggestion for developing this type of work, the PCNs refer to students carrying out investigations according to their interests and reality, enabling the data collected to be analysed, described, organised and interpreted more easily, as it has more concrete meaning for the students. They also point out that these procedures are often used to solve problems and that they encourage students to ask questions, establish relationships and build justifications, thus developing a spirit of critical enquiry and participating in the construction of mathematics itself.

According to Fainguelernent (1999, p.15), "from this perspective, it is necessary to make time available for the problematisation of knowledge, enabling the meaningful exploration of this information, contributing to the development of desirable competences for the formation of students as citizens, in the opportunity of a critical and investigative mathematical education. Awakening students' interest in the reality that surrounds them and the events they experience as citizens".

Polya (1887-1985), cited by Wachiliski 2007 p.31 as the father of Problem Solving as a methodology for teaching Mathematics, with the main objective of teaching students to think, says: **"his methodological technique is** very **close to** one of the trends in Mathematics education, which is **Mathematical** Modelling**".**

Computers are already part of our daily lives, at home, at work, in transport, at school, and in the hands of young people. The interactive environment makes computer-based learning fun and exciting. It generates a new engagement with learning and brings about new challenges, new ideas, new ways of constructing knowledge and developing thinking and leads to a revitalisation of educational debates.

In order to attract the attention of our students, and also so that we don't disappoint them, therefore: in order to have good results in maths lessons, the gap between the lesson and the reality of the student's life needs to be minimal, and of course, the desire to learn the right thing needs to be satisfied and reality needs to be satisfied.

According to Bessanezi, "Mathematical modelling is the art of transforming situations from reality into mathematical problems, the solutions to which must be interpreted in the usual language and **approximation of reality."**

We no longer live without interactive and synchronised communication and information devices, but how many know or know how programmes and applications really work. How many would know how to use a computer in DOS, the operating system that lit the fuse, changed the world's thinking and opened up the path to communication. It's true that in order to do maths, we still have to give up modern resources and spend so much time and energy on historical, uninteresting and already very tedious learning, with students rushing out of class to devices with calculators, many of whom have never used them and don't even know how to do so.

Knowing this, or perhaps counting on it, many could take advantage of the situation for their own benefit, and this is where we learn maths.

CHAPTER 2

MATHEMATICS

There is strong evidence that man has been on Earth for longer than we realised. At least 4 million years ago. As homo habilis began to organise his thoughts based on reflections, the use of memory and observations, he needed to explain and predict certain problems of understanding quantities. About 2 million years ago, he began to find ways of solving them, and the problems eventually gave rise to mathematical development. Man learnt to observe the numerical value by differentiating one from two and so on. Over time, by representing quantities through objects such as pebbles, he had the chance to put them together and so was able to interpret addition and subtraction. The first tools that helped in the interpretation and development of maths date back to this time.

Very old mathematical objects, such as the Ishango bone found in Zaire, today's Congo, and the Lebombo bone discovered in the Libombos mountains in Swaziland and dated to around 35,000 years B.C. Also the Lebombo bone which contains 29 notches made in a fibula, or fibula, of a baboon. Prehistoric artefacts have also been discovered in Africa and France, dating from between 35,000 and 2,200 years ago, which suggest archaic attempts to quantify time.

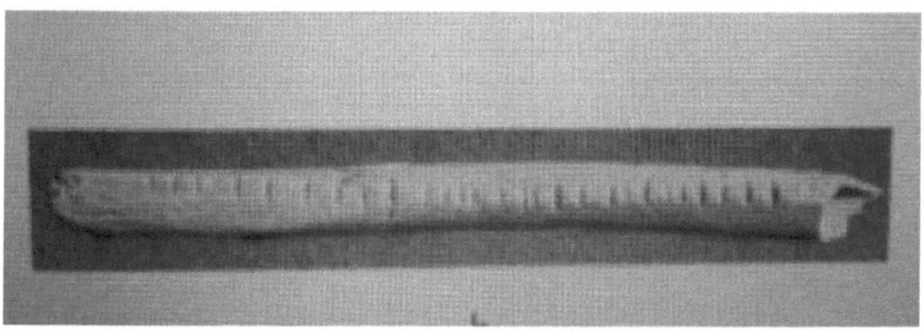

Ishango's bone

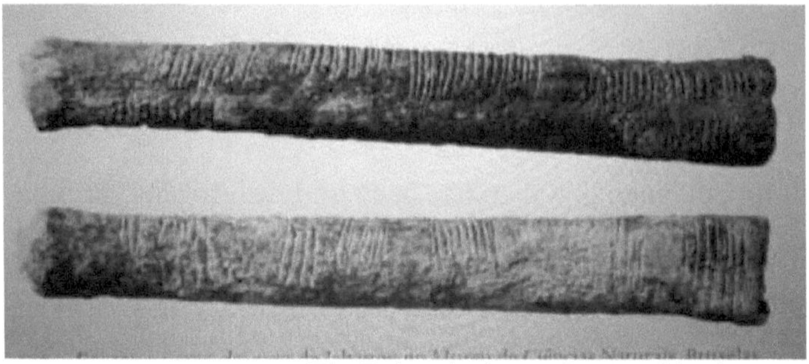

The Lebombo bone, front and back

In China, the abacus was invented, the first mechanical calculating instrument, and then tables and area calculation were created. These things happened between three thousand and two thousand five hundred years BC.

The Chinese abacus

The most archaic mathematical texts available are the Plimpton of Babylonian mathematics, circa 1900 BC, and the Rhind Mathematical Papyrus of Egyptian mathematics, circa 2000-1800 BC, as well as the Moscow Mathematical Papyrus **of** Egyptian mathematics, circa 1890 BC. All of these texts deal with the so-called Pythagorean Theorem, a mathematical advance that was widely disseminated after basic arithmetic and geometry. The Greek-Hellenic contribution greatly refined methods, especially through the introduction of deductive reasoning and mathematical rigour under proof, and expanded the subject of mathematics, i.e. what it deals with.

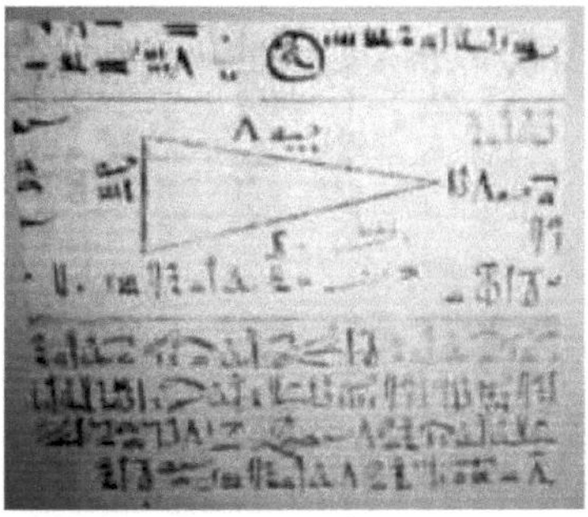 

Rhind Mathematical Papyrus
Plimpton's maths texts

The study of maths as a topic in itself began in the 6th century BC with the Pythagoreans, who coined the term "mathematics" from the Ancient Greek mathema, meaning "subject of enlightenment". Chinese maths made contributions very early on, including the positional notation system. The Indo-Arabic number system and the rules for using its operations, currently in use around the world, was probably developed around 1000 AD in India and transmitted to the West through Islamic mathematics. Islamic mathematics, in turn, developed and expanded the mathematics known to these civilisations. Many Greek and Arabic texts on maths were then translated into Latin, which contributed to the development of maths in medieval Europe.

From ancient times to the Middle Ages, the outbreak of mathematical creativity was often followed by centuries of stagnation. Beginning in the Renaissance, in the 16th century, new advances in mathematics, interacting with new scientific discoveries, were realised incrementally, continuing to this day.

The cognitive thinking of maths is certainly the beginning of the scientific spirit, as it has been proven naturally and throughout its use, it is the basis of various other activities in the exact sciences, such as engineering, architecture, chemistry, physics and many others that follow.

It was maths that led to the study of matrix calculus, which is the basis of the logical functioning of computer processors, and also to the study of vector calculus, which builds the diagrams that make up the screens and optical mice of devices.Maths is the only tool that makes it possible to know if an account closes correctly or if there is an error in its presentation. It is by having absolute mastery of it that many can take advantage of others who have not mastered it.

It is therefore still a compulsory part of the school curriculum.

Our ancestors Homo Hábilis

CHAPTER 3

TECHNOLOGY IN SCHOOLS

As needs have arisen, man has used his rational capacity to develop new technologies and mechanisms. Advances in technology are conceptualised as everything that leads someone's knowledge to evolve, improve or simplify their activities, or better still: it is any process of cultural improvement. Humanity has gone through several phases of technological evolution, but a common misconception when thinking about technology is that it refers to the latest generation.

Technology is the knowledge developed by an entire people, everything that they have built up over time and use in their lives for well-being, safety and evolution.

When it comes to information and communication, technological possibilities have emerged as an alternative, facilitating education through digital inclusion, with the insertion of computers in schools, allowing students to use technology, access information, and carry out multiple tasks in all dimensions of human life, as well as training teachers through the creation of networks and virtual communities, so it is a very useful tool.

The content linked to information processing, as described in the National Curriculum Parameters, is extremely important for our students' education and for familiarising them with technology.

According to Dolores Folador (2011 p. 122), "this is fundamental knowledge for reading the world, for understanding more complex texts, especially those that include information represented in graphs and tables".

Thus, the calculator and the computer, representatives of man-made technological growth, are part of what may yet come to complete human evolution, helping to support education and prepare students so that they can fulfil their role in society.

From the point of view of learning: the use of a calculator, as well as other technology, as Dolores (2011 p. 20) comments, "does not rule out the need for students not to reason in order to solve or formulate problems, but it does change the way in which they reason, leading them to responsibly use something other than pencil and paper".

It is worth mentioning studies such as Garcia and Borba's (1999), "interesting to note that those who support the use of calculators emphasise problem-solving rather than just calculating". The calculator does not diminish the user's ability and need to reason and work on their mathematical logic.

It is well known then that the use of technology in the teaching of maths is interesting, as Lopez (2003) points out, "by using calculators, the user is replacing traditional mathematical algorithms, accounts, with another calculation mechanism, which makes it possible to obtain faster results and also to compare these results more quickly".

The stage is set, many programmes are prepared to get our minds moving. Everything leads us to do maths, look at tables and analyse graphs, use data and plot results. Checking and comparing these results, criticising, pointing out errors, giving suggestions and solutions.

The need to use this knowledge exists because it will help prepare and learn about the new citizen, which is specifically the work of the educator. It is in fact the result of all the work of the teacher educator to prepare individuals who make up the entire population of a nation, and not to allow these individuals to be stripped of their native rights as people in a synchronised world.

It's clear that information and communication technology is growing alongside people's daily lives, and that the number of users of devices with internet access is increasing all the time. The fact that it is used by young people makes them, as we know, expert in the use of the machine, but not in real information. Knowing what to access, depending on the technology and its maintenance, is the domain and responsibility of the educator.

The computer is a resource that children like, but it is necessary to monitor its use critically so as to avoid exaggeration and damage to their education. A new way of learning is opening up.

It creates **the possibility for students to learn by "playing",** building their own knowledge, thinking and solving their mistakes in a constructive way. Technology stimulates the creative process, problem-solving and improves the human condition.

Various programmes by state and municipal education departments and the Ministry of Education (MEC) have brought computers and internet access to Brazilian public schools, but simply introducing these technologies is not enough to improve the quality of education.

Gimenez, quoting the GentequeEduca website, which can be accessed at https://gentequeeduca.wordpress.com/**, comments that "in** most cases, computers have arrived in schools without the backing of a pedagogical proposal". Gimenez **comments:** "The importance of play in early childhood education is essential and necessary, as it helps in the construction of identity, in the formation of individuals and in their ability to communicate with each other, reproducing their daily lives and characterising the learning process.

Despite the **'Technological Age'**, it is very important for children's development that they have contact with toys and access to games according to the region in which they live. This contact allows children to create their identity through a broader vision, because it is through social interaction that they will learn to deal with different types of **situations".**

The study by Mazzilli (2005), still on the GentequeEduca website, says: "It is necessary to go beyond the simple technical implementation of computers and the Internet in schools, and to understand how didactic-pedagogical relationships, involving students, teachers and managers, take place and what difficulties there are in these relationships".

It is necessary to delve deeper into ways of observing and analysing how interactions, practices, new possibilities and didactic-pedagogical relationships take place with the presence of new technologies.

An important step in this direction is to diagnose the current situation regarding the use of computers and the Internet in schools in order to identify successes and problems, as well as to point out ways forward. Concerned about this point and commissioned by the Victor Civita Foundation - FVC, a quantitative survey carried out by the Laboratory of Integrated Systems - LSI, with the participation of IBOPE INTELIGÊNCIA, investigates the use of computers and the internet in 400 public primary and secondary schools in Brazilian capitals.

Research has shown that the mere existence of computers in schools does not translate into improved school performance, although access to computers and the internet is highly valued by society and has a high impact. It recommends changing the focus of training programmes, both at undergraduate level and in continuing education, from learning how to use ICTs to learning how to learn using ICTs, and including technology in lesson planning and projects, socialising good practices among teachers, reflecting on new educational paradigms with the use of ICTs and different models of using and making computers available.

Here we can explain something about this subject. Information and

Communication Technologies - ICTs - correspond to all the technologies that interfere with and mediate the information and communication processes of human beings. They can also be understood as a set of technological resources that are integrated with each other and that provide, through hardware, software and telecommunications functions, the automation and communication of business processes, scientific research and teaching and learning. They enable participatory teaching with a view to reality.

Example of Information and Communication Technologies - ICTs

Obviously, equipping schools with computers, improving access to the internet and training teachers and students to use computers are important actions to promote and democratise access to the information that is essential for understanding the world around us.

Quoted on the same GentequeEduca website, Castro (2010) says, **"it remains to be seen how the use of computers can actually make a difference to learning".** It's important to remember that without a specific function, any machine loses its usefulness.

On the other hand, we have the observation that we are setting up a system, and it is being set up for patient users who will use what is made available to them, with a view to a better and more dignified future for everyone.

Talita Moretto, on the website sala aberta,

https://salaaberta.com.br/opiniao-a- informatica-as-tics-e-a-educacao/, comments: recently in 2016, "I **decided to take a "look" at the National** High School **Curriculum Parameters**, the Pcnem", an official document of the Ministry of Education - MEC, written in the year 2000, **on "Languages, Codes and Technologies",** this is where we find the defence of informatics in schools, and there it is also clear that the competences and skills to be developed concern the need to bring informatics closer to the socio-cultural life of students, relating the computer to real and everyday cases. Page 62 reads**: "recognise IT as a tool for new** learning **strategies** capable of making a significant contribution to the process of **building knowledge in the various areas".**

Once again, we recognise the intention to integrate students, technology and reality into teaching, in schools, during lessons.

Moretto continues, "The main question that arose for me was to investigate whether the introduction of computer science would really be able to contribute to a change in the attitude of the teacher who, when encouraged to use a computer, would realise that the instrument can provide opportunities for the creation and promotion of collaborative educational projects, stimulating student participation in the teaching and learning process, or whether it would just be an inanimate machine within the **school."**

It can thus be seen that the concern with having electronic devices or ICTs now lies in how to use them, and the answer to the question: what could these machines be used for?

On the other hand, there is a whole series of compulsory entries required by law and drawn up by our chosen ones, so that they can **"access"** the entries, known as transparency, and that's not all. These are huge numbers full of decimals, commas and amounts that reach into the billions, where a little mistake can easily result in millions.

This data makes us think about adding, subtracting, comparing, checking, constructing all kinds of graphs and commenting. Working with these numbers requires some attention, but it can be very interesting for those who are learning and for those who are teaching maths.

CHAPTER 4

INFORMATION PROCESSING: MATHEMATICAL DATA

Using technological means to learn and teach maths requires a certain amount of planning in terms of values. The search for the values to be worked on mathematically must attract the student's attention, bringing them into reality, teaching, and preparing them for the world in which they live, and the one they will pass on to future generations who will follow them.

The big problem here, as we already know, is making use of basic mathematical operations and being able to understand or interpret them, knowing their applications in cases close to reality and current problems, and helping human evolution. The history of maths basically tells us that it should bridge the gap between understandings, relationships, business and accountability, and lead everyone to a good mathematical understanding.

The search for knowledge and the use of Transparency is visibly everywhere, whether we are aware of it or not. The alternative of knowing and controlling the amounts, values and figures released and made available by public bodies, as required by transparency laws, is required.

In a more secure future, there can be no protection or favouring of error or the conscious benefiting of a few, at the expense of the majority. The reason for learning the art of maths and using the knowledge it teaches has precisely this as one of its main purposes.

Transparency portals, such as the Transparency Portal on the website of the Federal Government's Office of the Comptroller General - CGU, at http://www.cgu.gov.br/assuntos/transparencia-publica/portal-da-transparencia, for example, stipulate that every citizen must control the data they publish, in the following terms:

The Federal Government's Transparency Portal is a channel through which citizens can monitor the use of federal resources collected through taxes to provide public services to the population, as well as finding out about other matters related to the Federal Public Administration. The site, launched in November 2004, is maintained by the Ministry of Transparency, Supervision and Comptroller General of the Union - CGU, and aims to guarantee and expand transparency in public management and strengthen social participation in overseeing the spending and investments of the Federal Executive Branch.

The Portal provides information on federal public funds transferred by the Federal Government to states, municipalities, the Federal District, directly to citizens and to those in public office, as well as data on spending by the Federal Government itself on purchases or contracts for works and services. In addition, the space displays, with daily updates, documents from all phases of expenditure generated the previous day.

The tool also provides information on sanctions received by companies, non-profit organisations and individuals in their dealings with the Federal Public Administration and information on civil servants, including registration data, remuneration, use of functional properties and any punishments suffered, among others.

Of course, at first glance, this idea is no more frightening than a mathematical problem that asks for the ratio of how many fruits John has after winning, among all the kind contributors, having eaten several and donated some of them. Using data from Transparency is something that can become easy and interesting, especially if studied in a Mathematical Modelling exercise carried out in the classroom and with the whole class divided into groups of students.

The CGU website discreetly has a link below, like this:
Go to: www.transparencia.gov.br, which allows us to move on by opening the link, and initially we have a few options for queries. In the horizontal menu we can see the Expenses tab, which opens by default, and in this tab we have the options: "Expenses - Commitments, settlement and payment with daily update from 2010", and "Expenses - Payments, with monthly update".

On the "Expenditure" tab, we also have four tables that provide searches. These are: "Transfer of Funds", which allows us, in a drop-down menu, to search by financial year, in a period from 2004 to 2018, the options: "by
state/municipality", "by programme", "by action" and "by beneficiary", linked to the criteria **provided by another drop-down menu in the "Select" field, with the following** values: "government entities", "non-profit entities", **"other legal entities", "natural persons", and "legal entities by economic activities".**

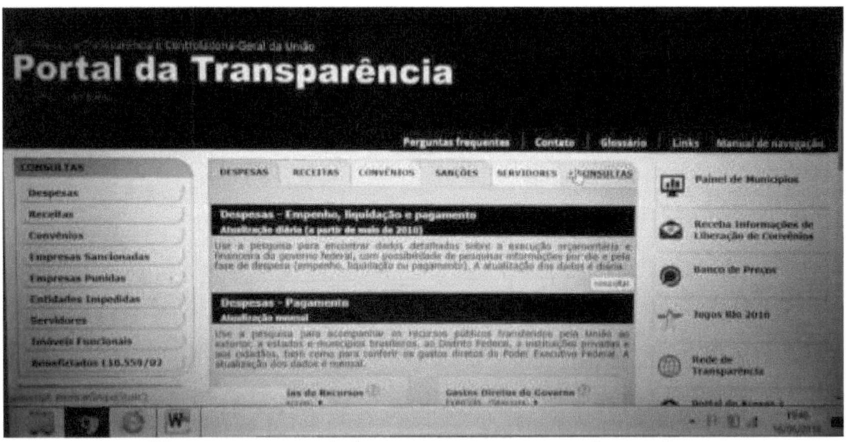

Transparency website at www.transparencia.gov.br

In the other "Direct Government Expenditure" table, you again

have the option, in a drop-down menu, of the year of the financial year in the same period as the other table, the **selection options: "by type of expenditure", "by executing body", "by programme", "by action" and "by beneficiary", with the following possible criteria in the "select" box: "government entities", "non-profit entities", "individuals, companies and others", and "legal entities by economic activity".** The two other boxes that **make up the "Expenditure" tab on the CGU Transparency website are: "thematic enquiries" which covers the"** family grant", "daily allowances paid", and the "federal government payment cards" with the options "by body" or "by bearer", in the financial years 2002 to 2018 **and a link "other thematic enquiries",** and the table "Enquiries by **budget function",** the area of action and purpose of the execution of Federal Government expenditure, through the sum of direct spending and transfers of **resources, with the enquiry options, "by function or area", and "by sub-function or purpose" in the financial** years between 2013 and 2008.

Other search options are easily found in the other tabs above: Revenues, Agreements, Sanctioned Companies, Entities
Impeded, Servants, Functional Property. For each of these options there is a
relative table that allows you to search by period.

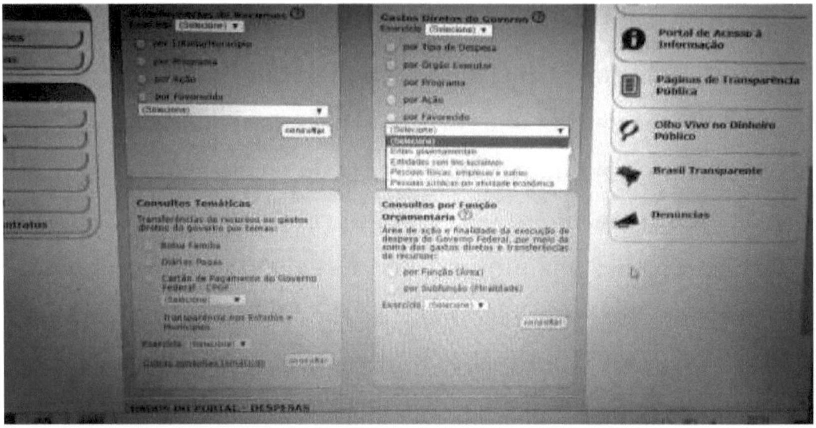

The CGU's Transparency in spending options board

If, for example, the "Expenses" option is chosen, the return is a page with various expenses listed by: category, origin, budget data, posting and breakdown. Each of these expenses is linked to a more detailed view in "Details", **which leads to a visualisation of the expense and opens the "By Line" option for detailing. In this case, the explanation divides the expense into several items, which makes it possible to continue the consultation in the "Sub-item" option, which lists some** expenses without forecasts and realised expenses.

It is interesting to note the characteristic of the entries and the values for mathematical calculations, related in divisions and subdivisions, each expense is divided into others and then into more others, making it possible to use these entries in Problem Solving projects involving Maths.

The resources presented by area refer only to the consolidation by budget function of the amounts transferred by the Federal Government to the states and finally to the municipalities, according to the classification of expenditure in the Integrated Financial Administration System - SIAFI, and do not reflect the total expenditure of the Federal Government.
Federal Government in these areas. Some options lead the researcher to

obtain the amounts passed on to the municipality in Reais, and this is divided according to some classifications. These include health, education and other related areas. The understanding of these values, purely of mathematical interest, turns on the operational use of calculus and a certain attention to functional realism.

Ideally, the same figures posted by the Union should be found directly on the respective websites of the public bodies, which in our example here would be that of Guaratuba City Hall in the state of Paraná - Brazil.

The visible contribution is in relation to the possible assembly of graphs and their interpretation and the percentage values of simple mathematical calculations.

It is worth quoting FOLADOR (2011 pg 101). "Students can begin to understand how statistical data is collected, organised and reported. It's not the number of graphs they put together that helps them build concepts, but how meaningful each of the activities carried out with these resources was."

The rational understanding of the mathematical drawings we've already discussed must didactically transpose that curiosity and adoration of colours and shapes into the intelligent and necessary understanding of the longed-for functional citizen.

The use of mathematical values of interest to the Mathematical Modelling project, provided by Transparency, becomes very pedagogical, and it is in the national interest for the values posted to be accessed and checked by the population. Values in figures, tables and graphs can be found in releases from any public body, such as town halls, city councils, state governments and others, which can be searched specifically by the name of the body of interest, through the corresponding websites. FOLADOR;

(2011, p. 104), refers to the use of these drawings **"more important than collecting the data and transforming it** into a **graphic** representation**, is interpreting the results in a creative way".** A study aimed at the use and understanding of numbers and drawings brings the idea of the Mathematical Modelling project, where after learning how to access the entries, check them through calculations, assemble colourful graphs, they can discuss and comment on them.

CHAPTER 5

MATHEMATICAL MODELLING

Maths is an important part of a citizen's education and should make its contribution to the extent that methodologies are explored that develop the student's intellectual capacities, structure of thought, speeding up logical and deductive reasoning, in the application of problems, in everyday life situations and in the activities of the world of work, as suggested by the PCNs.

The PCNs propose educational practices that are appropriate to the social, political, economic and cultural needs of the Brazilian reality, taking into account the interests and motivations of the student and guaranteeing the learning necessary for a citizen's education. They also point out that education should prioritise the contextualisation of content, give meaning to and encourage discussions around themes of relevance or cross-cutting themes, using different languages as a means of producing, expressing and communicating ideas in order to achieve these objectives. It suggests flexibility in the curriculum to adapt it to the different realities and needs of Brazilian education.

In order to adapt the teaching of mathematics to this current context, mathematical modelling and problem solving have been presented as ways of achieving these objectives. According to Bassanezi (2002, p.16), "Mathematical **modelling** is the art of transforming problems from reality into mathematical problems and solving them by interpreting their solutions in the language of the **real** world"**. These problems come from the student's own interests and the** mathematical **content** covered is generated from the theme to be problematised. The search for a solution is carried out by means of a Mathematical Model, which represents that situation, and a consequent validation, which involves verifying the solution found. As MACHADO (2006) comments, "In a learning environment with Mathematical Modelling, the teacher's situation is that of a mediator, work guide and, what we consider most important, a learner. In this environment, the teacher doesn't know everything, he or she will also learn and grow with each piece of work proposed. This is a situation in which the teacher will have to learn to cope, the insecurity of not having control in their hands; it is important to be open and to value the students' knowledge, because it is in this interaction that learning will take place."

Several authors (BARBOSA, 2001; BASSANEZI, 2002; BIMBENGUT; HEIN, 2003; BIEMBENGUT, 2004; BEAN, 2001; BURACK, 1994; SCHEFFER; CAMPAGNOLLO, 1998) as a methodology capable of bringing real-life situations into the classroom, transforming this science into an instrument so that students can interpret reality mathematically, acquiring skills both to move better in their environment and to intervene in it.

For Bassanezi (2004, p. 17), **"Mathematical modelling is a process** that combines theory and practice, motivating the user to seek an understanding of the reality that surrounds them in the search for ways to act on it and transform it". The link between the use of technology to support the classroom through the use of computers, combined with the pedagogical and methodological approach to mathematical knowledge, can be made in mathematical modelling, **which, for Barbosa, "is a** learning **environment in** which students are invited to problematise and investigate situations through mathematics, with references to **reality".**

Mathematical modelling, as Biembengut and Hein **(2002) state, "is**

the process that involves obtaining a model, so that a set of mathematical symbols and relationships try to translate, in some way, a **phenomenon in question or a problem in a real situation".**

There is a large body of literature on mathematical modelling, where many sources teach and deal with the methodology of the modelling project, its steps, divisions, stages, form of assessments and presentations, and also demonstrate in detail the role of the organiser in problematising and choosing themes.

In studies cited by Ribeiro (2008 p. 68 and 69), he reports that "experience shows that working on mathematical knowledge in a Mathematical Modelling project requires first of all complying with the steps in the sequence for constructing this activity".

Once the technology requirements, such as computers, internet and pupils, are in place, the steps must follow the organisation starting with the definition and selection of the Curriculum Content to be used or worked on, then follows:

- **o** Choosing the generating theme: a theme or area of reality whose curricular content will be studied;
- **o** Definition of the main question: this guides the treatment of the generating theme, i.e. it defines what the theme is intended to achieve;
- **o** Problematisation and Problem Solving: **the** phase for answering the Matrix Question based on the Problematisation and investigation of the Theme. This is the moment when Mathematical Knowledge emerges from the need to solve the Matrix Question;
- **o** Building Mathematical Knowledge: This stage is carried out concurrently with Problematisation and Problem Solving. It guarantees the construction of concepts as problems are solved;

- Solution of the problematised situation; time for discussion, evaluation and analysis of the solutions obtained;
- Presentation: communication of the results achieved (fairs, posters, exhibitions, experience reports, Power Point, etc.);
- Retrospect: critical reflection seminar on the project developed.

The Mathematical Concepts are those that we must pass on to the students, which in this article, as I mentioned before, are the basic mathematical operations, but any other concept from the Mathematical Curriculum can be used to develop the Mathematical Modelling activity to take advantage of Transparency in Public Administration.

CHAPTER 6

METHODOLOGY - PRACTICAL OVERVIEW: EXAMPLES

Example 1 - Mathematical Modelling Project.

The transparency of the federal government's transfers to the state of Paraná, based on the month of November 2014, for example, could be used to make a mathematical modelling game possible.

Data source for calculations: The data posted on the Transparency Portal of the Federal Government's Office of the Comptroller General, at http://pr.transparencia.gov.br/receitas/por-area/areas?exercicio=2014, provides us with mathematical values that can be treated mathematically.

We realised that during the course of this work, the email address above, which provided us with the data for this first example, was removed and that the data presented in recent searches for the same period in question is not close to that published here. The figures only apply.

The modelling methodology has already been explained in the previous chapter:

The steps in the construction sequence for this activity are as follows.

the mathematical concept to be worked on, the four basic mathematical operations (addition, subtraction, multiplication and division).

o **Generating Theme**: Study of Transparency in Federal Government transfers to the state of Paraná, base month: 11/2014.

o **Matrix question** adopted: the importance of each area in the total monthly value of the transfer and in the total annual value, check the values found.

o **Problematisation:** for this Matrix, it is the division of the value of the related area by the total monthly value and the total value for the year, it is the sector's share of the total monthly and annual transfer respectively and the inverse operation of verification.

o **Building Mathematical Concepts:** when finding the contribution, both monthly and annual, the concept used is division and the result will be a decimal rational number. When checking, we will use the concept defined as sum and multiplication, then fill in the proposed table.

o **Solution:** assemble the suggested table and present the Mathematical Concepts work in writing.

The following items refer to the **Presentation** and the **Retrospective,** which should be inherent to each work/study group and also to the assessment system defined by the project's supervising teacher.

Area	Total 11/2014	Total per year
ALL	299.460.138,54	3.362.636.780,43
Special Charges	289.539.371,36	3.065.307.802,29
Education	330.875,89	110.039.885,17
Citizenship Rights	0,00	51.185.522,19
Industry	6.596.615,66	41.061.101,30
Health	335.559,10	19.850.967,45
Labour	0,00	17.481.799,80
Urban planning	0,00	12.099.424,01
Social Assistance	0,00	12.092.400,00
Science and Technology	2.190.796,53	12.058.648,57
Public Safety	0,00	6.052.248,94
Sports and Leisure	0,00	4.606.659,80
Agriculture	0,00	4.381.263,00
Agrarian Organisation	0,00	2.218.406,80
Commerce and Services	466.920,00	1.942.071,32
Housing	0,00	1.467.052,80
Sanitation	0,00	520.476,43
Administration	0,00	271.050,56

Table 1 data 11/2014 - PR

As for the table suggested for modelling, it could take the form of the table above, which shows the values copied from the Portal for the month of November 2014.

Problematisation requires the sum of the values of all the accounts specified in each row, which must close at the value above and be titled "All", both for the month in question, the first column, and for the annual value, when the second column is added.

Another problematisation may be required in this same project, and will depend on the mathematical concepts to be dealt with, such as percentages, for example, or the cumulative value of the increase in each bill.

Drawings, such as graphics, can be included in the game, and must be demonstrated, provided there is interest in animating the project.

Example 2 - Mathematical Modelling Project.

The total allocated by the Federal Government at national level (TdpGFan), from 2014 onwards, has an account entitled "Primary Health Care Programme (pABS)", and this has a sub-account entitled "Construction and Expansion of Basic Health Units (caUBS)". Let's tabulate these national transfer values, respectively between 2014 and the current 06/2018*.

We can calculate the percentages of each value in relation to the total transfer made for the corresponding annual period.

From the Federal Government's Transparency Portal at http://www.transparencia.gov.br/, in the first box, entitled "Transfer of **Resources", we'll select the option "By programme" and in the "select" field we'll choose the value "**Government **Entities" from the drop-down menu, for all the years in the required period, on the page that follows,**

we'll choose the account "Basic Health Care" which, **when accessed, will give access to the Link entitled** "Construction and Expansion of Basic Health Units" **which, when accessed, makes it possible to** choose by Tax Units or States of the Union.

Notice that three accounts are shown above, in the header of the site.

It is these accounts that we will be working on in this Mathematical Modelling project using Transparency.

The accounts for each base year in the period are:

Base year	TdpGFan R$	%	pABS R$	%	caUBS R$	%
2015	339.744.202.337, 16	100	84.855.825,76		49.143.549,95	
2016	369.408.894.501, 04	100	54.124.715,37		29.421.649,90	
2017	365.445.385.247, 32	100	13.832.065,71		5.715.549,90	
2018 *	128.583.900.286, 92	100	1.386.588,42		648.083,34	

Table 2 Data Transfer to the Health Programme, 2014 - 2018 *

The steps in the construction sequence for this activity are as follows.

the **Mathematical Concept** to be worked on, The calculation of percentages.

- o **Generating Theme**: Study of Transparency in the Federal Government's transfer to the Unified Health Programme for the construction and expansion of Basic Health Units, from 2014 to 2018 inclusive (*).
- o **Matrix question** adopted: the annual balance of the transfer relating to each allocation to the Generator Programme, the importance of each account.
- o **Problematisation:** for this Matrix, it is the division of the value of the account or sub-account by the Total value of the previous account of the Grouping Programme in each base year.
- o **Construction of Mathematical Concepts:** find the annual

percentage of each account and sub-account of the transfer to the Basic Health Programme. The concept used is division and the result will be a decimal rational number that multiplied by one hundred expresses the percentage value. In the verification, we will use the concept defined as division and multiplication, then fill in the proposed table.

- **Solution:** assemble the suggested table and present the Mathematical Concepts work in writing.

Below are the items relating to the **Presentation** and the **Retrospective,** which must be inherent to each work/study group and also to the assessment system defined according to the criteria chosen by the project's supervising teacher.

Data source for calculation work:

The data posted on the Transparency Portal of the Federal Government's Office of the Comptroller General.

Example 3 - Mathematical Modelling Project.

The amount allocated by the Federal Government under the name of Bolsa Família can be analysed from the website of the Transparency Portal of the Ministry of Transparency and the Attorney General's Office, where you have the vertical yellow menu on the **left, selecting the button "Portal in Graphics", in** the first box **called "Bolsa Família",** when accessed it will open in a new tab to the email address:
http://www.portaltransparencia.gov.br/graficos/bolsafamilia/. At
On this page, we can see vertical bar graphs of the amounts transferred and the number of benefits. We must select a Fiscal Unit, for example the State of São Paulo and the Municipality of São Paulo, as well as São Vicente. Using the values shown on the R$ Values and Number of Benefits tabs, we

will construct the following table according to the years from 2009 to 2018* (the current period).

Municipality	São Paulo		St Vincent	
Year	Value R$	No. Benefits	Value R$	N⁰ Benefits
2009	178.935.002	239.763	11.434.561	15.969
2010	130.626.369	249.637	11.676.634	15.523
2011	229.373.417	232.196	12.446.576	12.852
2012	272.976.486	263.731	12.815.552	11.350
2013	430.243.182	415.386	14.090.624	11.216
2014	588.403.288	461.627	13.329.842	9.798
2015	718.026.906	538.938	14.466.046	9.464
2016	799.838.006	596.585	14.720.111	9.973
2017	824.765.545	624.919	17.699.727	11.111
2018	146.832.843	517.054	3.255.137	9.644

Table 3 Bolsa Família Programme transfer data, 2009 - 2018*

With the values tabulated above, we can add up the lines referring to the amounts allocated in Reais and obtain the total unadjusted values for the municipality for the period between 2009 and 2018*, or in another way, by obtaining the inflation indices for the respective years and then multiplying the annual values by the respective indices, we will obtain the real value of the adjusted amount for the same period. Now let's think about dividing each base year by the respective number of beneficiaries, and we'll get the annual percentage value. Adding up the annual per capita values and dividing them by the number of base years will give us the average per capita transfer value for the period between 2009 and 2018*, which can also be adjusted by the inflation index with the respective base years.

In this example, students will have to learn how to set up tables, do addition, multiplication and division.

The figures obtained are unbelievable. These figures should be analysed, not least because the launches don't always coincide with the figures published by the media, from where we learn about them. How much can you imagine the percapita value of Bolsa Família in Brazil is? You can do a simple and quick division between two figures obtained from the Transparency website and find out.

Oh, and you can use any technological support you like to do these maths operations.

The steps in the construction sequence for this activity are as follows.

the mathematical concept to be worked on, the four basic mathematical operations (addition, subtraction, multiplication and division), and the calculation of percentages.

o **Generating Theme**: Study of Transparency in the Federal Government's transfer to the Bolsa Família Programme in Value and Number of Beneficiaries, from 2009 to 2018 inclusive (*).

o **Matrix question** adopted: verification of the total annual amount transferred to the municipality in the period under analysis, corrected and not corrected, the percapita amount per base year and the average percapita amount over the same period, transferred to each municipality analysed.

o **Problematisation:** for this Matrix, it is the sum of the value of all the transfers to each Municipality without correction and corrected when multiplied by the respective annual inflation indices. Obtaining the values per beneficiary for each base year analysed for each municipality observed and this average value over the period studied.

o **Construction of mathematical concepts:** find the total unadjusted value of the transfer per municipality analysed, and this after adjustment by the respective annual adjustment indices. Calculate the per capita or per beneficiary values for each base year and the per capita value for the period and per municipality.

o **Solution:** assembly of the tables suggesting the values obtained plus the results found, and written presentation of the Mathematical Concepts work.

What follows are the items relating to the **Presentation** and the **Retrospective,** which must be inherent to each work/study group and also to the evaluation system defined according to the criteria chosen by the teacher guiding the project, as defined in the final part of the Chapter entitled: Mathematical Modelling.

Example 4 - Mathematical Modelling Project.

The amount allocated by the Federal Government under the heading of Transfer by Locality can be analysed from the website of the Transparency Portal of the Ministry of Transparency and the Attorney General's Office, where **you have the vertical yellow menu on the left, selecting the button "Portal in Graphics", and in the** second **box called** "Transfer by Locality", you can access the link: "Illustrated consultation of information on transfers of resources made by the Federal Government" and a new tab will open to the e-mail address: http://www.portaltransparencia.gov.br/graficos/transferenciasporlocalidade/. On this page, you can see graphs in the form of vertical bars that refer to the amounts transferred by base year, and next to them, by region, which you can select in the graph in the form of a multicoloured circle, each colour referring to a region of the country. The site's engine has simple HTML with few search features, and the only accesses apart from the regions in the coloured circle **are two buttons above with the options "Overview" and "UF". In this example, the** Mathematical Modelling **game** we're going to play is the construction of graphs using the amounts transferred by the Federal Government to the Southeast region in the base year of 2018. Each working group can work with different data.

Selecting the green region in the multicoloured circular graph opens a table showing the four states in this region: ES, RJ, MG, SP. So let's assemble the table and see the data for the current base year 2018.

UF	TRANSFER R$
ES	1.476.408.732,74
MG	5.426.320.182,91
RJ	5.747.929.685,76
SP	7.452.421.953,03

plotting these values on graphs.

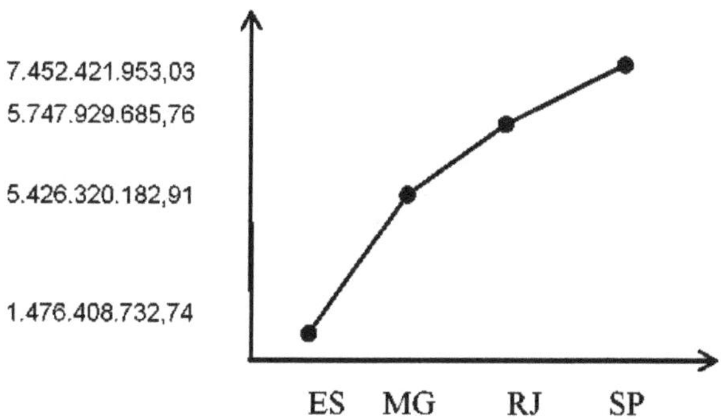

Line graph plotting table values

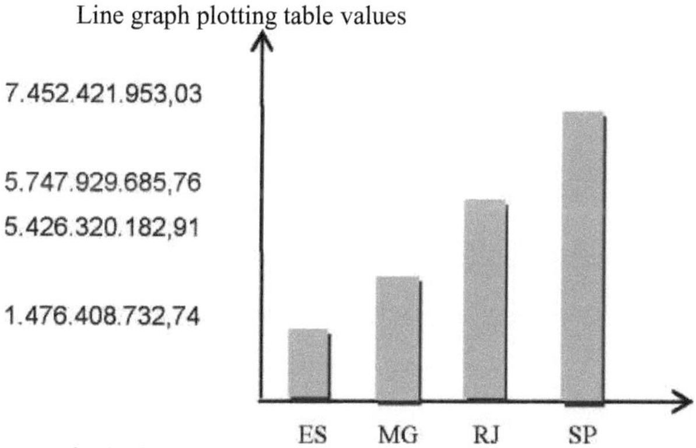

Bar graph plotting the table values

In these graphs we can understand the magnitudes of the values transferred, whether they are: one in relation to the other, in relation to the null value, and in relation to the average references represented. It is essential to interpret graphs consciously, interpreting the ordinates and coordinates and the values they represent. Reading graphs should be understood like the letters of the alphabet or even mathematical numbers, leaving no doubt. This is how the steps in the construction sequence for this activity are.

the **Mathematical Concept** to be worked on, Construction and Interpretation of Graphs.

o **Generating Theme**: Study of Transparency in Transfers from the Federal Government to States and Municipalities in the Southeast Region, in Real Values, in the current base year of 2018 (*).

o **Matrix question** adopted: Verification of the amounts transferred to each state in the region in a line graph and a horizontal bar graph for 2018.

o **Problematisation:** for this Matrix, each value must be in orders of magnitude, with reference to the axis that defines them and linked to the axis that makes up each FU studied.

o **Building Mathematical Concepts:** Draw the table of transfers on graphs showing the orders of magnitude, in relation to each other and relative to the axes and the origin or absolute zero of these axes. The values plotted should be clear and easy to understand.

o **Solution:** assemble the tables, choose the type of graph that represents the table and provides a quick interpretation of the values studied

This is followed by the items relating to the **Presentation** and the **Retrospective,** which must be inherent to each work/study group and also to the assessment system defined according to the criteria chosen by the teacher guiding the project, which are defined in the final part of the Chapter entitled: Mathematical Modelling.

CHAPTER 7

RESULTS

Maths teaching goes back millennia and seeks to transmit knowledge that prepares students for a life of knowledge. Forming individuals who are prepared for social life is the idea behind the reform of maths curricula, as it is in the interest of the students' participatory training in the reality of teaching and learning that brings solutions to social problems, with a view to a prosperous and culturally evolved nation. The maintenance and growth of a people's culture in relation to their own interests is therefore of great importance, because our students of today will be the citizens of tomorrow and it is good that they know, use and even teach maths.

The opportunity is opening up with the arrival of information technology, which is bringing new paths for these new citizens, as well as knowledge of their interests in relation to the world in which they live and understanding that one day they will have the responsibility of teaching and preparing those who will follow.

FINAL CONSIDERATIONS

We believe that in order to train the brand new citizen, there is a need for greater dedication to study, requiring grants and qualitative and participatory research projects in this area of mathematics, at postgraduate level such as Masters and Doctorate, open to the student and/or graduate population in conjunction with the sources that release the data, considering the need and the government's strong interest in transparency.

We therefore feel that it is essential to maintain controlled development, with the opening of new university courses and new professional ideas to overcome the new difficulties encountered. The idea of the sacred path of humanity proposes new solutions, as opposed to accepting defeat or creating opportunists who fight and hide the solution to social problems.

It is of the utmost importance that Mathematical Knowledge be present in the formation of new societies, because it is the essence of the knowledge of this science that solves most of today's problems. The mediating aid of mathematics, whether in education and the formation of new citizens, or in support of the already existing and active bodies that control or release data, is required, and gives us the opportunity to do so by solving problems in mathematical modelling in school education.

BIBLIOGRAPHICAL REFERENCES

BARBOSA, J. C. **Mathematical modelling: conceptions and experiences of** *future teachers.* 2001. 253 f. Thesis (Doctorate) - Institute of Geosciences and Exact Sciences, Paulista State University, Rio Claro, 2001.

BASSANEZI, R.C. *Teaching and learning with mathematical modelling - a new strategy*. São Paulo: Contexto, 2004.

BASSANEZI, R.C. *Teaching and learning with mathematical modelling - a new strategy*. São Paulo: Contexto, 2002.

BIEMBENGUT, M.S.. *Mathematical modelling and its implications for the teaching and learning of mathematics*, Blumenau SC, FURB 1999.

BRAZIL, Ministry of Education, Secretariat for Primary Education, National Curriculum Parameters: Mathematics, Brasilia MEC/September 2008

_____ Parâmetros Curriculares Nacionais de Matemática para o Ensino Fundamental. Brasília: MEC / SEF, 1998.

_____ Secretariat of Basic Education. National curriculum parameters:

Mathematics/ Secretariat of Fundamental Education - Brasília: MEC/ SEF, 1998. 148p.

Available at: http://www.bibvirt.futuro.usp.br/textos/hum

_____ , Ministry of Education and Sports. *Education Guidelines and Bases Law*

promulgated on 20 December 1996. São Paulo, Ed. do Brazil, 1996.

____ Ministry of Education. Department of Basic Education. **Parameters**

National Curriculum: Maths. Brasília, 1997.

____ . **National Curriculum Parameters:** third and fourth cycles. Presentation of

Cross-cutting themes. Brasilia, 1998.

____ . **Curriculum proposal for youth and adult education:** second segment of the programme

Ensino Fundamental: 5a a 8a Série: Introdução/Secretaria de Educação Fundamental. Brasília,

2002.

CHAVES, E.O.C.; The Computer in Education, 05/2014, Works done, in:

http://www.trabalhosfeitos.com/ensaios/Computador-Na-School/53037932.html

DORIA M.; proposta de informática tic tic tac, 2014, at: http://www.tictictac.com.br/proposta de informática tic tic tac 2014.pdf

ECXELEÊNCIAS; Transparência Brasil, at http://www.excelencias.org.br/index.php, accessed on 01/2015.

FAINGUELERNENT, E. K. **Educação Matemática**: Representação e construção em geometria, Porto Alegre. Artes Médicas Sul.

FOLADOR, Dolores; **Special Topics in Maths Teaching; Technologies and Information Processing**. Curitiba. IBEPEX 2011.

GARCIAS. T.A.S.; BORBA, M.C., Graphing Calculators and Quadratic Functions**.** In: BORBA M.C. ; FAINGUELERNENT, E. K.; GOTTLIEB, F.C.; (Org). Graphing Calculators and Maths Education, 2ª ed. Rio de Janeiro, Art Bureau, 1999,. Pg. 59-74.

GENTEQUEEDUCA The use of computers and the internet in public schools at http://www.fvc.org.br/estudos-e-pesquisas/avulsas/estudos1-7-uso- computadores.shtml, accessed on 12/2014

INSTITUTO PAULO MONTENEGRO; National Institute of Functional Literacy, at http://www.ipm.org.br/ipmb_pagina.php?mpg=4.02.01.00.00&ver=por accessed 01/2015.

LOPEZ, A. J.; **Exploring the use of calculators in teaching maths to young people and adults.** Available at: http://www.matematicahoje.com.br/telas/autor/artigos/artigos view.asp?cod=37

MACHADO E. S.; Mathematical Modelling and Problem Solving Problems, Dissertation presented for a Master's degree, by the Postgraduate Programme in Science and Mathematics Education at PUC-RS, on 01/2015

MORAN, J. M. **New Challenges in Education:** The Internet in face-to-face and virtual education. Available at " http//www.eca.usp.br/prof/moran/novos.htm".

MUDA MAIS, 16 July 2014, in: 01/2015. http://mudamais.com/daqui-pra-melhor/99-das-escolas-publicas-brasileiras- has-computers

PORTAL DA TRANSPARÊNCIA; Comptroller General of the Union, Federal Government at http://www.portaltransparencia.gov.br/ access, on 01/2015.

REVISTA ESCOLA, 07/2014, Tecnologia na Educação, at:01/2015 **http://revistaescola.abril.com.br/blogs/tecnologia-educacao/2014/07/16/cresce-uso- de-tablets-e-computadores-na-escola/**

RIBEIRO, F. D. **Games and Modelling in Mathematics Education:** Curitiba IBEPEX. 2008.

TAJRA, S. F. **Information in Education:** New pedagogical tools for today's teacher. São Paulo. Ética 2001.

WACHILINSKI M; DIDÁTICA DA AVALIAÇÃO, ALGUMAS Perspectivas da Educação Matemática, IBPEX, 2008.

More Books!

yes I want morebooks!

Buy your books fast and straightforward online - at one of world's fastest growing online book stores! Environmentally sound due to Print-on-Demand technologies.

Buy your books online at

www.morebooks.shop

Kaufen Sie Ihre Bücher schnell und unkompliziert online – auf einer der am schnellsten wachsenden Buchhandelsplattformen weltweit! Dank Print-On-Demand umwelt- und ressourcenschonend produzi ert.

Bücher schneller online kaufen

www.morebooks.shop

info@omniscriptum.com
www.omniscriptum.com

OMNIScriptum

MIX
Papier aus verantwortungsvollen Quellen
Paper from responsible sources
FSC® C105338

Printed by Books on Demand GmbH, Norderstedt / Germany